BYEGIG

Ahead of AI

An Operator's Guide to Daily Use and the Infrastructure Behind It

A practical guide for technology-literate readers who want to use artificial intelligence confidently, understand the systems beneath it, and prepare for what comes next.

FIRST EDITION

ByeGig | July 2026

Product features, laws, standards, and market conditions change quickly. Readers should verify current information and apply professional judgment, especially in legal, medical, financial, security, and other high-stakes settings.

Published by ByeGig

byegig.com

FRONT MATTER

Preface

Why this book begins with daily practice and ends with infrastructure.

Artificial intelligence is becoming ordinary

Artificial intelligence is moving through the same transition that transformed the personal computer, the internet, and the smartphone. It is becoming less visible as a novelty and more important as a layer of everyday activity. People increasingly use AI to draft messages, understand unfamiliar subjects, organize information, compare alternatives, analyze documents, write code, prepare meetings, and make sense of complex decisions. The most useful applications are often not dramatic. They are small improvements repeated across hundreds of ordinary tasks. [1]

This shift is occurring unusually quickly. Stanford's 2026 AI Index estimated that generative AI reached roughly 53 percent population-level adoption within three years, faster than the early adoption curves of the personal computer or the internet. OECD data show that the share of firms reporting AI use rose from 8.7 percent in 2023 to 20.2 percent in 2025. Adoption remains uneven, but the direction is clear: AI is becoming part of how knowledge work is performed rather than a separate technical specialty. [2]

The two levels of AI literacy

Most introductory guides stop at the interface. They explain how to ask a chatbot a question, draft an email, or summarize a document. Those skills matter, but they are only the first level of literacy.

The second level begins when a user asks what sits behind the interface. Why does an AI system sometimes sound certain when it is wrong? What is the difference between a model, an application, an agent, and a workflow? What are context, memory, retrieval, tools, and permissions? Why do data centers, electric grids, semiconductors, identity systems, financial settlement networks, robots, satellites, and national policy increasingly belong in the same conversation?

ByeGig's subject-matter focus is the infrastructure beneath emerging technologies. This book therefore moves in two directions at once. It teaches readers how to use AI now, and it introduces the systems that will determine how AI scales. The first half is practical. The second half is architectural and strategic. Together they form a more durable kind of literacy.

BOOK THESIS

The central idea

AI literacy is not just knowing how to generate an answer. It is knowing how to define a task, supply the right context, verify the result, protect sensitive information, and understand the infrastructure that makes the result possible.

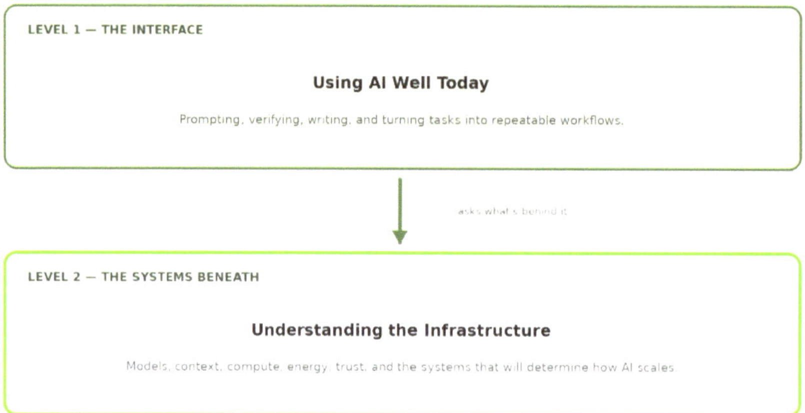

Figure: the two levels of AI literacy this book builds — daily practice, and the infrastructure beneath it.

FRONT MATTER

How to Use This Book

Begin with the task in front of you, then deepen your understanding one layer at a time.

A progressive structure

1. Part I builds daily competence: asking better questions, researching, writing, analyzing, and turning isolated tasks into repeatable workflows.
2. Part II explains the intermediate systems behind those capabilities: models, context, retrieval, agents, compute, trust, robotics, finance, sovereignty, and networks.
3. Part III turns knowledge into practice through a personal adoption system, team governance principles, and a 30-day implementation plan.

Every chapter includes an introductory layer and an intermediate layer. The introductory layer answers: What is this, why does it matter, and how can I use it today? The intermediate layer answers: What is happening behind the interface, what can go wrong, and how does the subject connect to the larger AI economy?

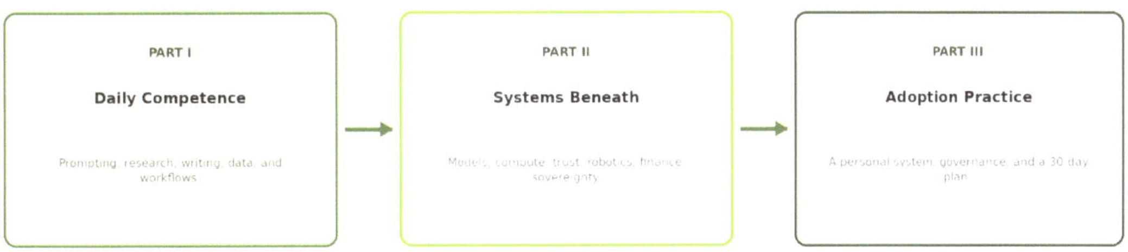

Figure: how the three parts of this book connect — daily competence, the systems beneath it, and the adoption practice that ties them together.

A rule for reading

Do not try to memorize every term. Use the book to build a mental map. The details will change. The enduring concepts are task definition, context, evidence, evaluation, permissions, coordination, infrastructure, and trust.

The most productive way to learn AI is to alternate between use and reflection. Apply one concept to real work. Observe where the system helps and where it fails. Then return to the underlying idea. Practical experience makes the architecture easier to understand; architecture makes practical experience safer and more effective.

FRONT MATTER

Contents

Part I

AI as a Daily Tool

The first objective is not technical mastery. It is dependable usefulness.

CHAPTER 1

AI Is Becoming an Everyday Layer

The most important change is not that AI can do extraordinary things. It is that it can assist with ordinary things continuously.

From product to layer

A new technology usually starts out as a destination: people visit a website, open a specialized program, or set aside time to use an unfamiliar tool. Mature technologies, though, tend to fade into the background and become layers instead — search became a layer of browsing, GPS became a layer of transportation, and cloud services became a layer beneath the software people use every day. AI appears to be heading the same way.

For the individual user, this means the most valuable question is no longer, 'What can AI do?' The better question is, 'Where in my day do I repeatedly translate, compare, summarize, structure, draft, calculate, explain, or decide?' Those are the places where AI can become an operating layer rather than an occasional experiment.

The shift is visible in workplace behavior. OpenAI reported in January 2026 that more than a quarter of U.S. workers, and 45 percent of workers with postgraduate degrees, said they used ChatGPT for work. The pattern matters more than the brand: AI tools are becoming a first step in many workflows, especially where a person must move from an unstructured problem to a structured next action. [3]

The four everyday roles

- Explainer: translate a difficult concept into a level, analogy, or format you can use.
- Editor: improve clarity, structure, tone, and completeness while preserving your intent.
- Analyst: compare alternatives, organize evidence, identify assumptions, and expose missing information.
- Simulator: role-play a customer, interviewer, critic, regulator, buyer, or decision-maker so you can test an idea before the real interaction.

These roles are deliberately modest — none of them require the system to act as an oracle. Instead, they treat it as a flexible cognitive instrument, while the user stays responsible for goals, judgment, evidence, and action.

Use it today

Choose one recurring task that consumes between ten and forty-five minutes. Good candidates include preparing a meeting, rewriting a difficult email, comparing a contract clause, turning notes into an agenda, summarizing an industry development, or outlining a decision. Run the task with AI three times over one week. Save the prompt, note what changed, and record how much review was required.

The purpose is not to prove that AI is universally useful. It is to identify the exact conditions under which it creates leverage for you. A useful AI practice is built from verified personal workflows, not generic lists of possibilities.

COPY-AND-USE PROMPT

Find your highest-value starting points

Review the following description of my typical workday. Identify ten recurring tasks where AI could reduce friction without requiring the AI to make the final decision. Rank them by likely time saved, risk, and ease of adoption. For the top three, propose a simple before-and-after workflow.

My workday: [paste a description]

The intermediate layer: augmentation versus automation

AI use exists on a spectrum. At one end, the system suggests language while the user controls every decision. In the middle, the system completes a defined task and the user reviews the result. At the far end, an automated process takes actions across tools with limited intervention.

The farther a workflow moves toward automation, the more important permissions, monitoring, audit trails, data classification, exception handling, and rollback become. The same task that is low risk as a draft can become high risk when connected to an email account, payment system, customer database, or physical machine. Practical AI literacy therefore includes knowing when not to automate.

SAFETY PRINCIPLE

Start with reversible work

Early AI workflows should be easy to inspect and easy to undo. Drafting, summarizing, classifying, brainstorming, and preparing options are safer starting points than sending, purchasing, approving, diagnosing, or executing.

CHAPTER 2

What Modern AI Actually Does

A useful mental model for large language models, multimodal systems, and probabilistic output—without pretending they think exactly like people.

Prediction at enormous scale

Modern generative AI systems learn patterns from large collections of data and use those patterns to produce new outputs. A language model predicts useful continuations of text. A multimodal model can relate patterns across text, images, audio, and other forms of information. A reasoning-oriented system may spend additional computation exploring candidate steps before answering. None of these descriptions captures every technical detail, but together they provide a practical model: the system generates plausible, context-sensitive responses rather than retrieving a single guaranteed truth from a database.

This explains both the strength and the weakness of these systems. The same flexibility that lets a model draft a strategy, explain a legal concept, write code, or imitate a style also lets it produce statements that sound coherent without actually being accurate. Fluency is one capability; truthfulness is a separate requirement, and a model can have plenty of the first without much of the second.

The five concepts every user should know

- Model: the trained system that transforms input into output.
- Prompt: the instruction and context supplied for a task.
- Context: the information currently available to the model, including the conversation, uploaded files, tool results, and system instructions.
- Tool: an external capability such as web search, code execution, a database, calendar access, or an application interface.
- Evaluation: the process used to decide whether the output is good enough for its purpose.

Many weak AI workflows focus on the prompt alone, but stronger ones manage all five concepts at once — they use an appropriate model, supply relevant context, connect tools when current or precise information is needed, and evaluate the result against explicit criteria.

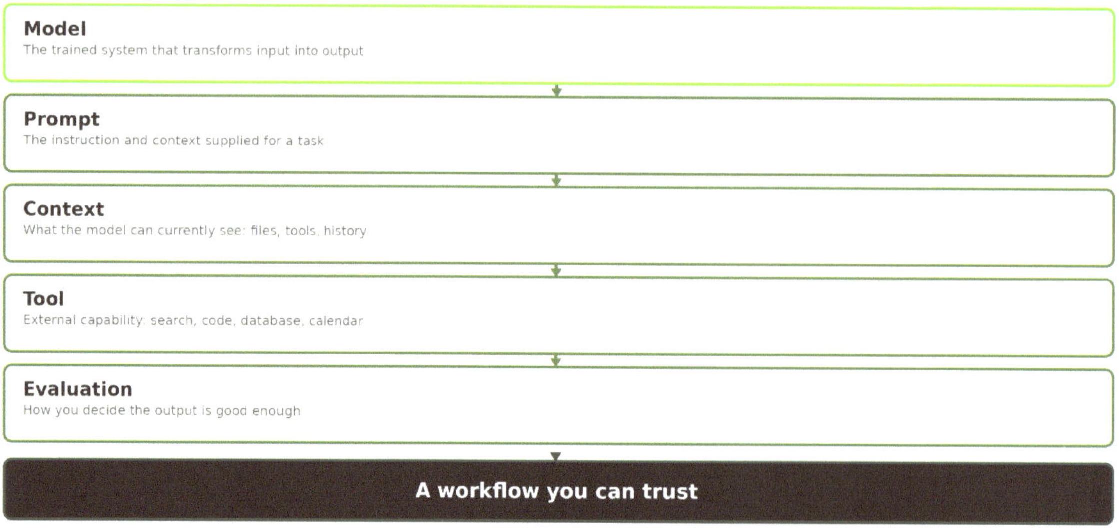

Capability is task-dependent

It is misleading to ask whether an AI system is simply 'smart.' Performance varies by task, context, tools, and evaluation method. A system may produce an excellent explanation of a concept and a poor estimate of a market price. It may write syntactically correct code that fails in an unusual environment. It may summarize a provided document accurately but invent facts when asked about a current event without access to reliable sources.

Treat capability as a matrix rather than a ranking. Ask: What kind of input is available? Does the task require current facts? Is there a correct answer? Can the output be tested? What is the cost of being wrong? What human expertise is required to review it?

CORE CONCEPT

A practical definition

An AI system is reliable enough when the task is well defined, the necessary context is available, the output can be evaluated, and the consequences of error are controlled. Reliability is not a permanent property of the model; it is a property of the entire workflow.

The intermediate layer: training, inference, and adaptation

Training is the resource-intensive process through which a model learns statistical patterns. Inference is the process of using the trained model to generate an output. Adaptation methods—including instructions, retrieval, tool use, fine-tuning, and feedback—shape how a general model performs a specific job.

For most users, the distinction matters because it clarifies where improvement can come from. A disappointing result does not always require a more powerful model. It may require better context, a clearer output format, a relevant source, an example of success, or a method for checking the answer. Prompting is one part of system design, not magic wording.

Practice exercise

Ask the same AI system to complete one task under three conditions: first with a one-sentence request, then with detailed context and an output format, and finally with the same context plus an example of a strong result. Compare accuracy, relevance, and the amount of editing required. The exercise demonstrates why context and evaluation often matter more than clever phrasing.

CHAPTER 3

Building a Daily AI Practice

Move from random questions to repeatable, inspectable workflows that compound over time.

The problem with occasional use

Many people use AI intensely for a day and then forget about it. They ask a few broad questions, get uneven results, and conclude that the technology is either magical or overrated — and neither conclusion does much for them in the long run.

A daily practice is different. It connects AI to recurring work, uses consistent inputs, saves successful instructions, and improves through review. The goal is not maximum usage. The goal is reliable placement: knowing which moments benefit from AI and which do not.

The daily loop

1. Capture: collect the notes, documents, questions, or data that define the task.
2. Frame: state the objective, audience, constraints, and definition of success.
3. Generate: ask for options, analysis, a draft, or a structured transformation.
4. Inspect: verify evidence, assumptions, calculations, tone, and completeness.
5. Revise: provide feedback or make edits yourself.
6. Reuse: save the prompt, template, checklist, or workflow if the task will recur.

The last step is where the compounding returns come from: a useful prompt becomes a template, a template becomes a workflow, and a workflow can eventually grow into a shared tool or an agent.

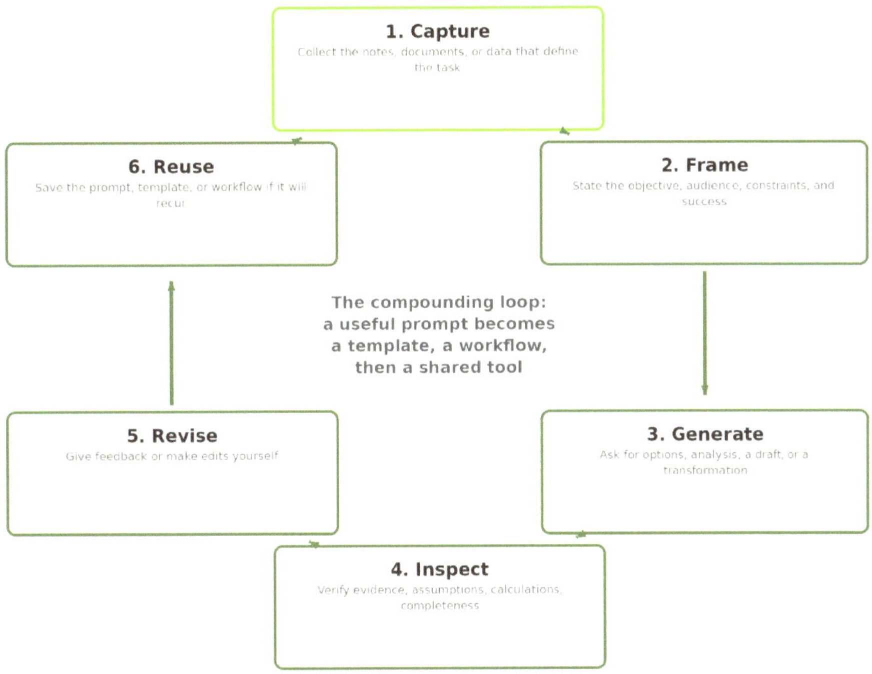

Three practical routines

Morning orientation: provide your priorities, calendar constraints, and open questions. Ask the system to identify dependencies, likely bottlenecks, and the most important decision of the day. Do not ask it to choose your priorities without context; ask it to help you see the structure.

Midday compression: turn accumulated notes, messages, and documents into a concise status update. Ask what remains unresolved and what information would change the decision.

End-of-day capture: summarize what happened, what was learned, and what must carry forward. This can reduce the cost of restarting complex work the next morning.

COPY-AND-USE PROMPT

Daily operating brief

Act as my chief of staff for this planning session. Using the priorities, meetings, deadlines, and unresolved issues below, produce: (1) the three outcomes that matter most today, (2) dependencies and risks, (3) the questions I should answer before noon, and (4) a realistic sequence of work. Do not add commitments I did not provide.

Inputs: [paste]

The intermediate layer: personal context systems

As usage becomes more sophisticated, the limiting factor often shifts from model capability to context management. A system cannot reliably help with a long-running project if every conversation begins from zero. Personal context systems solve this by organizing goals, reference documents, decisions, terminology, preferences, and open questions so they can be supplied when needed.

This does not require giving an AI tool unrestricted access to everything. In fact, indiscriminate access creates privacy and relevance problems. A better approach is deliberate context: a project brief, a decision log, a glossary, a set of approved sources, and a current-status file. The user curates the working memory.

NON-NEGOTIABLE

The privacy boundary

Do not place confidential, regulated, proprietary, or personally sensitive information into an AI system unless you understand the product's data controls and your organization authorizes the use. Redact, abstract, or substitute data when possible.

CHAPTER 4

Prompting as Clear Delegation

The best prompts resemble good briefs: they define the work, provide the necessary material, and make success observable.

Prompting is not incantation

Prompting is often presented as a collection of secret phrases. That framing is appealing and mostly unhelpful. The durable skill is specification: communicating enough information for another capable party to perform a task, while preserving room for useful judgment.

Official guidance from major model providers converges on a simple idea: be clear about the task, provide relevant context, specify the desired output, and refine iteratively. The quality of the interaction depends less on theatrical language than on whether the system can understand what success looks like. [4]

The O-C-I-C-O-E framework

- Objective: What must be accomplished?
- Context: Why does the task exist, and who is the audience?
- Inputs: What source material, facts, examples, or data should be used?
- Constraints: What must be avoided, preserved, limited, or treated as uncertain?
- Output: What form should the result take?
- Evaluation: What criteria determine whether the result is acceptable?

A complete delegation prompt has six parts

O	C	I
Objective What must be accomplished?	**Context** Why does the task exist, and who is the audience?	**Inputs** What source material, facts, or data should be used?
C	**O**	**E**
Constraints What must be avoided, preserved, or treated as uncertain?	**Output** What form should the result take?	**Evaluation** What criteria determine whether it's acceptable?

Not every prompt needs all six labeled sections — think of the framework as a diagnostic tool instead. When an answer disappoints you, ask yourself which element was missing.

COPY-AND-USE PROMPT

A complete delegation prompt

Objective: Prepare a concise decision memo comparing [Option A] and [Option B].

Context: The audience is [role], who cares most about [priorities].

Inputs: Use only the facts and documents provided below.

Constraints: Identify uncertainty; do not invent prices, dates, or legal conclusions.

Output: One-page memo with recommendation, comparison table, risks, and next steps.

Evaluation: The recommendation must follow from the evidence and clearly state what would change it.

Materials: [paste or attach]

Use examples when form matters

If you care about tone, structure, level of detail, or formatting, an example is often more effective than a long abstract description. Show a strong prior memo, email, analysis, or table and ask the system to match its structural qualities without copying protected or inappropriate content.

Examples also make evaluation easier. They turn an ambiguous instruction such as 'make it executive' into observable characteristics: short paragraphs, explicit recommendation, quantified trade-offs, and a limited number of headings.

Iterate like an editor

The first output is a draft, not a verdict. Give feedback that is specific and causal: 'The recommendation is too confident because the cost estimate is unverified'; 'The introduction repeats the conclusion'; 'The table hides the key trade-off'; 'Preserve the technical detail but reduce jargon.' This teaches the system what to change and helps you clarify your own standards.

OpenAI's current prompt guidance explicitly recommends iterative refinement: start with a prompt, inspect the output, then add context, simplify, or adjust the instruction. That pattern is more reliable than attempting to write a perfect prompt in one pass. [4]

The intermediate layer: prompt chains and structured outputs

Complex work is often better handled in stages: first extract the facts, then identify the gaps, then compare the options, and only then draft the recommendation. Separating these stages reduces the chance that analysis and presentation get tangled together.

Structured outputs—tables, labeled fields, JSON, checklists, schemas, or fixed headings—make results easier to inspect and pass into later steps. This is the bridge from conversation to workflow design. Once output has a predictable structure, it can be reviewed systematically or used by another tool.

INTERMEDIATE PATTERN

A strong sequence

Extract → verify → analyze → challenge → present. Asking for all five at once can work, but separating them makes errors easier to find and the workflow easier to improve.

CHAPTER 5

Research, Verification, and Judgment

AI can accelerate research, but it cannot remove the obligation to distinguish evidence from inference.

Research begins with a claim map

Before searching, define what must be known. Break the question into claims: dates, quantities, definitions, causal explanations, current status, and competing interpretations. This prevents the research process from becoming a collection of interesting but irrelevant facts.

Ask the system to produce a claim map and mark each claim as stable, time-sensitive, disputed, or inferential. Stable claims can usually be supported by standard references, while time-sensitive ones need current sources. Disputed claims call for multiple credible perspectives, and inferential claims should be labeled as reasoning rather than presented as fact.

COPY-AND-USE PROMPT

Build a research plan

I need to answer this question: [question]. Break it into the smallest factual and analytical claims that must be established. For each claim, state: (1) whether it is stable or time-sensitive, (2) the best source type, (3) what evidence would be insufficient, and (4) how the claim affects the final conclusion.

Use a source hierarchy

1. Primary sources: laws, filings, official statistics, technical documentation, research papers, original datasets, and direct statements.
2. Authoritative synthesis: government agencies, standards bodies, recognized research institutions, and high-quality industry reports.
3. Reputable reporting: useful for events, context, and competing accounts, especially when primary material is incomplete.

4. Commentary and social sources: useful for leads, practitioner experience, and sentiment—but rarely sufficient for a central factual claim.

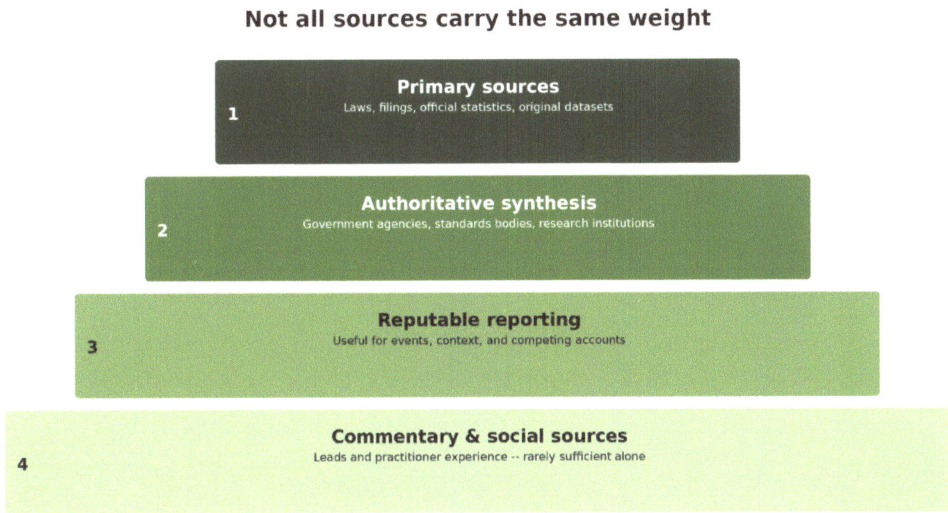

AI can help locate and summarize sources, but the user should inspect the source itself when precision matters — a citation alone isn't proof that the source actually supports the sentence attached to it. Check the relevant passage, date, methodology, and scope before relying on it.

The three-column method

For any important research product, maintain three columns: Evidence, Interpretation, and Decision. Evidence records what the source actually establishes, interpretation records what you think it means, and decision records what action follows from that. Keeping them separate prevents an elegant narrative from quietly turning assumptions into facts.

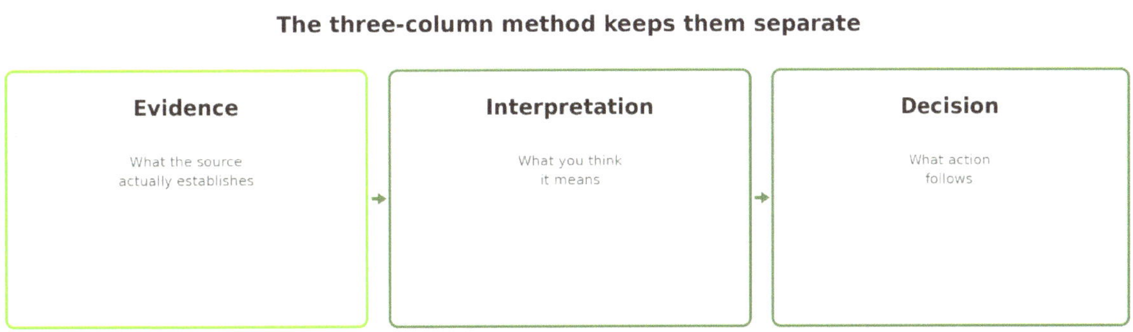

This method is especially valuable in emerging fields, where forecasts and promotional claims are common. The present may be observable while the future remains conditional. Use phrases such as 'the evidence suggests,' 'one plausible implication is,' and 'this would become more likely if' when the logic is inferential.

The intermediate layer: adversarial verification

A strong research workflow includes a deliberate challenge stage. Ask the system to identify the weakest evidence, alternative explanations, missing stakeholders, and conditions under which the conclusion fails. Then verify those criticisms independently.

This is not an invitation to create artificial balance. Some claims are much better supported than others. The purpose is to test the structure of your reasoning before another person does.

COPY-AND-USE PROMPT

Red-team an analysis

Act as a skeptical reviewer. Do not rewrite the analysis. Identify the five most consequential weaknesses in its evidence, assumptions, causal logic, or missing context. For each weakness, explain what additional source or test would materially strengthen the conclusion. Distinguish fatal problems from ordinary uncertainty.

Analysis: [paste]

JUDGMENT BOUNDARY

High-stakes rule

For medical, legal, financial, safety, regulatory, and security decisions, use AI to prepare questions, organize evidence, or draft material—not as the sole authority. Confirm with qualified professionals and current primary sources.

CHAPTER 6

Writing, Communication, and Creative Work

Use AI to improve expression without outsourcing authorship, judgment, or accountability.

The strongest use is transformation

AI is particularly effective at transforming material that already exists: turning notes into an outline, an outline into a draft, a long document into a briefing, technical language into a plain explanation, or a rough message into a professional communication.

Transformation is usually safer than generation from nothing because the source material anchors the result. The user can compare output against intent. It also preserves the most important creative input: the underlying idea.

A five-stage writing workflow

1. Discovery: clarify the audience, objective, argument, and evidence.
2. Architecture: create the sequence of ideas before polishing sentences.
3. Drafting: produce a complete version with explicit placeholders for missing facts.
4. Editing: improve clarity, rhythm, specificity, and economy.
5. Integrity review: check factual claims, citations, originality, tone, and whether the final position is truly yours.

A five-stage writing workflow

Different stages require different prompts. Asking the system to brainstorm and finalize simultaneously often produces generic writing because it commits to language before the idea is fully developed.

> COPY-AND-USE PROMPT
>
> ### Preserve the author's voice
>
> Edit the text below for clarity, structure, and concision while preserving the author's point of view and characteristic tone. Do not add new factual claims. Identify any sentence whose meaning is ambiguous before changing it. After the revision, provide a short list of the most consequential edits.
>
> Text: [paste]

AI as an audience simulator

One of the highest-value creative uses is perspective testing. Ask how a customer, investor, engineer, regulator, skeptical executive, or new employee might interpret the same material. The goal is not to accept every reaction. It is to identify assumptions that are invisible to the author.

This works especially well for product descriptions, proposals, presentations, public statements, and difficult emails. A simulated audience can expose jargon, missing context, unintended hostility, or a recommendation that arrives before the evidence.

The intermediate layer: style systems and content operations

At scale, organizations can turn writing preferences into a style system: approved terminology, examples, audience profiles, prohibited claims, citation rules, formatting standards, and review checkpoints. AI can then help enforce consistency without flattening every author into the same voice.

The intermediate challenge is governance. Who owns the source of truth? How are outdated claims removed? Which content requires legal or technical review? How is AI-generated material labeled or recorded? A content workflow is an information system, not merely a prompt.

AUTHORSHIP PRINCIPLE

A useful division of labor

Let AI accelerate structure, variation, editing, and audience testing. Keep humans responsible for the idea, evidence, decision, taste, and final accountability.

CHAPTER 7

Data, Documents, and Decisions

The practical leap occurs when AI moves from conversation into the materials where work actually lives.

Documents are structured problems in disguise

A contract, spreadsheet, policy manual, proposal, transcript, or operating report may look very different on the surface, but each contains recurring structures: entities, dates, obligations, exceptions, quantities, relationships, and unresolved questions. AI can help extract and organize those structures.

The safest first step is not 'analyze this document.' It is to define the schema: what fields should be extracted, how uncertainty should be represented, and what source location should accompany each answer.

> COPY-AND-USE PROMPT
>
> ### Extract with traceability
>
> From the attached document, extract the following fields into a table: [fields]. For every entry, include the page or section where it appears and a short supporting quotation. Use 'Not stated' rather than inferring a value. Add a final section listing ambiguities, conflicts, and information that would require human interpretation.

Spreadsheets and calculations

AI can explain formulas, propose a data-cleaning plan, generate code, classify rows, create scenarios, and help interpret results. It should not be trusted merely because the output is numerical. Validate formulas, units, date ranges, denominators, missing values, and whether a correlation is being mistaken for a cause.

A dependable workflow separates data preparation, calculation, and interpretation: inspect the source data first, then specify the transformation, then run or review the calculation, and only at the end ask what the result does and doesn't support.

Decision support, not decision theater

AI-generated matrices can create an illusion of rigor. A table with scores, weights, and decimals is not necessarily a good decision model. The important work is choosing criteria, justifying weights, identifying thresholds, and testing sensitivity.

Ask the system to show how the recommendation changes when assumptions change. If a small adjustment reverses the result, the decision is fragile and should be presented that way.

COPY-AND-USE PROMPT

Build a decision model

Help me compare the options below. First ask which criteria are truly decision-relevant. Then propose a transparent scoring model with definitions for each score, identify assumptions, and run a sensitivity test. Do not choose a winner until I approve the criteria and weights.

Options and known facts: [paste]

The intermediate layer: structured data and semantic retrieval

Organizations increasingly connect AI systems to document repositories, databases, and knowledge platforms. The system retrieves relevant material and uses it to answer a question. This is often called retrieval-augmented generation, or RAG.

The concept is simple; reliable implementation is not. Documents must be current, permissions must be respected, retrieval must find the right passages, and the answer must remain grounded in the retrieved evidence. Poor source hygiene produces confident access to outdated information.

Retrieval-augmented generation (RAG), simplified

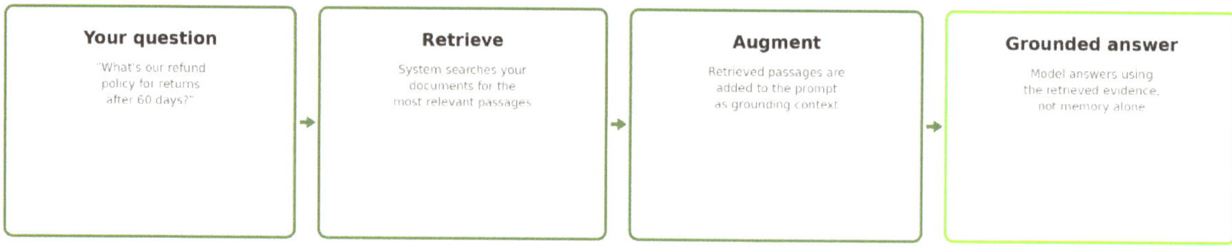

The concept is simple: reliable implementation depends on data quality, permissions, and retrieval accuracy.

INFRASTRUCTURE INSIGHT

Data quality becomes answer quality

When AI is connected to organizational knowledge, the model may be the most visible component, but version control, metadata, permissions, document ownership, and retrieval quality often determine whether the answer is useful.

CHAPTER 8

From Tasks to Workflows: Automation and Agents

The progression from asking for help to delegating a sequence of actions changes both the value and the risk.

A task, a workflow, and an agent

A task has a bounded input and output: summarize a report. A workflow connects tasks in a defined sequence: collect reports, extract metrics, compare changes, draft a briefing, and route it for review. An agent is a system that can pursue an objective across multiple steps, use tools, observe results, and decide what to do next within defined limits.

The word agent is used loosely. A useful operational definition is: a model plus tools, context, a control loop, and permissions. Without those surrounding components, an agent is only a conversational model with a more ambitious label.

The autonomy ladder

1. Suggest: propose a next step.
2. Prepare: create the material required for the next step.
3. Execute with approval: take the action after a human confirms.
4. Execute within limits: act independently inside predefined boundaries.
5. Manage exceptions: monitor a process, handle ordinary cases, and escalate unusual ones.

Most organizations should climb this ladder gradually. The system earns autonomy through evidence. A workflow that cannot be evaluated manually should not be automated blindly.

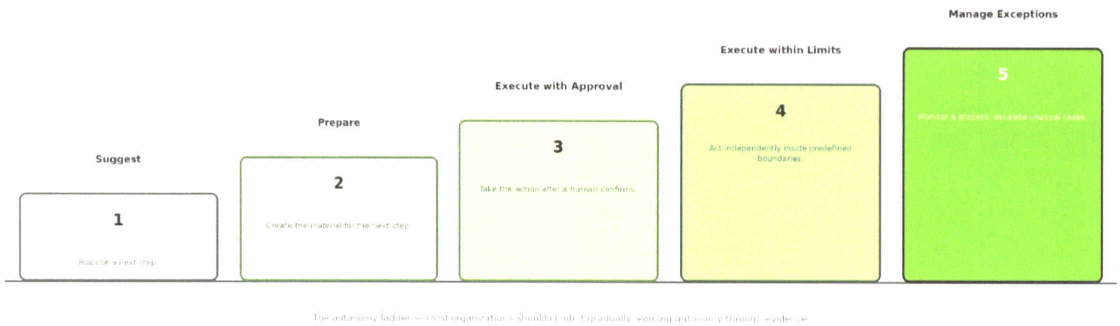

Figure: the autonomy ladder — from suggesting a next step to managing exceptions in a fully bounded workflow.

Design permissions before capability

A highly capable agent with broad access can make broad mistakes. Define what it may read, create, modify, send, purchase, approve, or delete. Set monetary, time, and scope limits. Require confirmation for irreversible actions. Record what tools were used and why.

This is the same logic used in cybersecurity and financial controls: authority should be limited to the minimum necessary for the job. AI does not eliminate the principle of least privilege; it makes it more important.

> COPY-AND-USE PROMPT
>
> ### Map a workflow before automating it
>
> Map the process below as a sequence of steps. For each step, identify: input, action, output, decision rule, required system access, failure modes, and whether human approval is needed. Then recommend the safest level on the autonomy ladder. Do not propose full automation if the process lacks measurable success criteria.
>
> Process: [describe]

The intermediate layer: orchestration and control planes

As organizations deploy multiple models and agents, coordination becomes a distinct infrastructure problem. Systems must route tasks, allocate compute, manage identities, track state, enforce policy, evaluate output, and resolve conflicts. This coordinating layer is often described with terms borrowed from cloud infrastructure: orchestration, runtime, control plane, and observability.

This is a central ByeGig thesis. The first wave of AI attention focused on models; the next wave increasingly focuses on the systems that make many models, tools, agents, and users work together reliably — because intelligence creates capability, but it's coordination that creates scale.

INTERMEDIATE CHECK

The agent test

Before calling a system an agent, ask four questions: What objective can it pursue? What tools can it use? What permissions does it have? How is its performance observed and stopped?

Part II

The Systems Beneath the Interface

Daily usefulness rests on models, infrastructure, energy, identity, institutions, and networks.

CHAPTER 9

Models, Context, Memory, and Retrieval

Intermediate AI literacy begins by separating the trained model from the temporary information and external systems surrounding it.

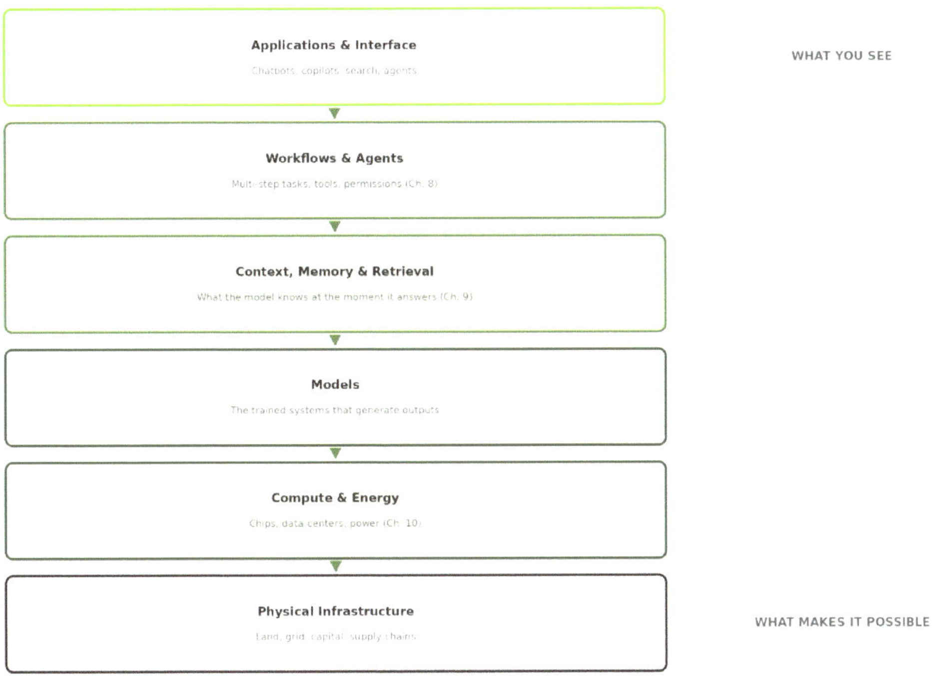

Figure: the AI stack — what a user sees at the top, and the layers of context, models, compute, energy, and physical infrastructure that make it possible.

The model is not the whole product

A consumer AI application combines many components: a model, interface, system instructions, safety policies, conversation history, file handling, search, code execution, memory, and connectors to external services. Two products using similar models can behave differently because the surrounding system is different.

This distinction is essential when evaluating tools. The question is not only, 'Which model is best?' It is also, 'Which system has the context, controls, integrations, and evaluation process required for this job?'

Context is working memory

The context window is the information available during a particular interaction. It may include your prompt, previous messages, attached documents, retrieved passages, and tool results. More context can help, but irrelevant or conflicting context can reduce performance.

Good context management is selective. Supply the material necessary for the task, identify which sources take priority, and separate background from instructions. When a project is long, maintain a concise project state rather than repeatedly pasting everything.

Memory is a design choice

AI products may remember preferences or project information across sessions, but memory should not be treated as perfect or unlimited. Important facts belong in controlled records. Use memory for convenience and explicit project documents for accountability.

Organizational memory adds further complexity. A useful system must know not only what information exists, but who is allowed to access it, whether it is current, and which version is authoritative.

Retrieval and grounding

Retrieval gives a model access to information it was not trained to reproduce reliably or that has changed since training. A system searches a collection, selects relevant passages, and uses them while generating an answer. This can reduce unsupported claims, but only when the retrieved material is relevant and the system cites or exposes its basis.

Grounding is the broader principle: constrain an answer to known material, tools, or rules. A grounded system should be able to say 'the source does not establish that' rather than filling a gap with plausible language.

The intermediate layer: embeddings, fine-tuning, and evaluations

Embeddings represent meaning in a numerical form that helps systems find conceptually similar material. Fine-tuning adjusts model behavior using examples. Evaluations test performance on a defined set of cases. These techniques solve different problems: retrieval supplies knowledge, fine-tuning shapes behavior, and evaluations measure whether the system works.

The most mature AI programs invest heavily in evaluations because improvement requires a stable definition of success. Without test cases, teams optimize by impression. With test cases, they can compare prompts, models, retrieval systems, and workflow changes systematically.

SYSTEM DESIGN

Do not confuse memory with truth

A system may remember a preference correctly and a fact incorrectly. Durable records, source links, and version ownership remain necessary even when the interface feels continuous and personal.

CHAPTER 10

Compute, Energy, and the Physical Infrastructure of AI

Every apparently weightless AI interaction depends on chips, data centers, networks, cooling, land, capital, and electricity.

The hidden physical system

AI is experienced as software, but advanced models are produced and operated through industrial infrastructure. Specialized chips perform enormous volumes of computation. Data centers supply power, cooling, networking, storage, and physical security. Fiber and high-speed interconnects move data. Semiconductor fabrication, packaging, and equipment supply chains determine how quickly capacity can expand.

This physical layer matters because it introduces scarcity. Software can be copied cheaply; high-performance compute cannot be expanded instantly. Capacity depends on factories, construction, grid connections, equipment, water, permits, talent, and capital.

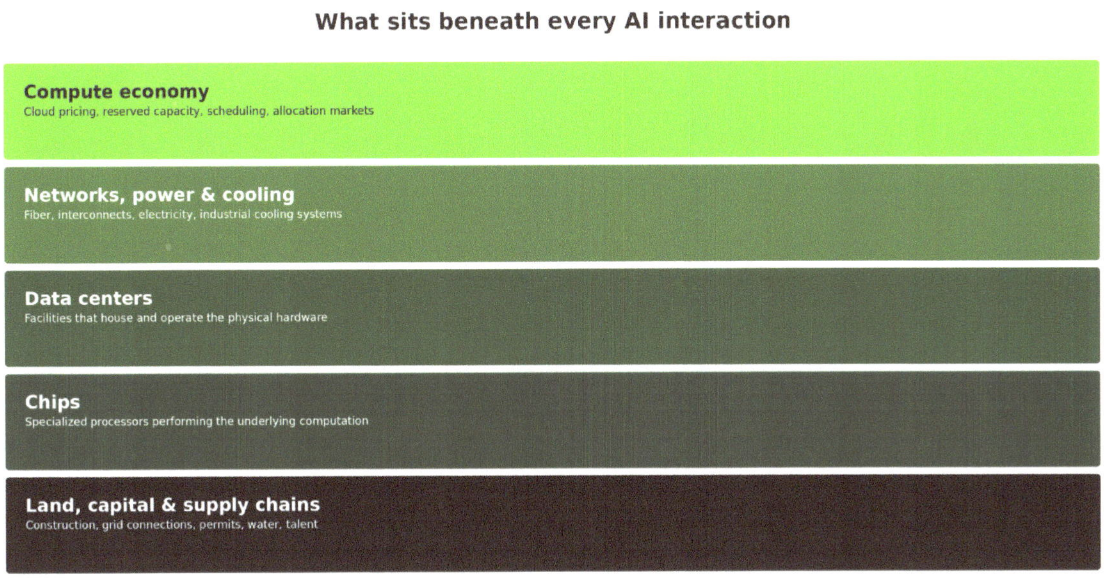

Energy becomes part of the AI stack

The International Energy Agency projects that global data-center electricity consumption could more than double to roughly 945 terawatt-hours by 2030 in its base case, with consumption growing around 15 percent per year from 2024 to 2030. The total remains a minority of global demand, but the local effects can be intense because large facilities concentrate load in particular regions. [5]

Why daily users should care

Infrastructure shapes availability, price, speed, privacy, and product design. A system running a large reasoning process costs more than a lightweight classification. A model operating on a device has different privacy and latency characteristics than one operating in a remote data center. A service that promises instant global access must manage capacity across regions.

Understanding this helps users interpret product trade-offs: faster isn't always better, larger isn't always necessary, and local isn't always more private if the device itself is insecure. The correct architecture just depends on the task.

The intermediate layer: compute as an economic resource

As compute becomes strategic, markets develop around allocation. Cloud pricing, reserved capacity, spot access, specialized accelerators, scheduling systems, and workload optimization are early forms of a broader compute economy. The underlying question resembles energy and logistics: how should a scarce, time-sensitive resource be priced, routed, reserved, and governed?

This is where concepts central to ByeGig's portfolio—compute exchanges, capacity markets, resource allocation, infrastructure operating systems, settlement, and sovereign compute—begin to connect. The names may evolve, but the coordination problem is real.

BYEGIG LENS

Infrastructure changes the question

At the application layer, the question is: What can the model do? At the infrastructure layer, the questions are: Where will the computation run, who controls access, what does it cost, how is capacity allocated, and what happens when demand exceeds supply?

CHAPTER 11

Trust, Identity, and the Verification Economy

When synthetic content becomes abundant, reliable provenance and authenticated participants become more valuable.

AI changes the economics of imitation

Images, voices, documents, and messages that once required specialized skill can increasingly be produced at low cost. The central problem is not that every synthetic artifact is harmful. The problem is that appearance alone becomes weaker evidence of origin.

This shifts value toward verification: confirming who or what produced an action, whether content has changed, what sources support a claim, and whether a person, machine, or agent has permission to act.

The everyday trust protocol

1. Pause when a message creates urgency, secrecy, or an unusual payment request.
2. Confirm identity through a second channel that you initiate independently.
3. Inspect the original source, not only a screenshot or forwarded excerpt.
4. Use unique credentials and strong authentication.
5. Record important approvals and decisions in systems with audit trails.

The everyday trust protocol

1. Pause when a message creates urgency, secrecy, or an unusual request
2. Confirm identity through a second channel you initiate independently
3. Inspect the original source, not just a screenshot or forwarded excerpt
4. Use unique credentials and strong authentication
5. Record important approvals and decisions in systems with audit trails

AI makes these familiar security habits more important. Verification should be procedural rather than emotional. A convincing voice or video is no longer sufficient proof.

Provenance and Content Credentials

The Coalition for Content Provenance and Authenticity develops the C2PA standard, which attaches verifiable provenance information to digital assets. Content Credentials are often described as a nutrition label for digital content: they can record information about creation and modification so that compatible systems can inspect the asset's history. [6]

Provenance is not a universal truth detector. Credentials can be absent, stripped, or misunderstood, and a properly credentialed item can still be misleading. The value is narrower and important: a standardized method for making claims about origin and history inspectable. The C2PA 2.4 specification was published in April 2026, reflecting the continued maturation of this infrastructure. [6]

The intermediate layer: machine identity

Agents and robots need identities just as users and services do. A system must know which agent made a request, what organization owns it, what tools it may use, how credentials are rotated, and who is accountable for its actions.

Machine identity becomes foundational when software can initiate transactions or control physical systems. The question 'Who are you?' expands into 'What are you, who authorized you, what can you do, and how can your actions be revoked?'

EMERGING CATEGORY

Trust is becoming infrastructure

In an AI-rich environment, trust cannot rely only on appearance or reputation. It increasingly depends on identity, provenance, permissions, secure channels, evidence, and auditable process.

CHAPTER 12

Physical AI, Robotics, and the Edge

Intelligence leaves the screen when systems can sense, decide, and act in the physical world.

From information to action

Generative AI produces information. Physical AI connects perception and reasoning to machines that move, manipulate, inspect, transport, or control real-world processes. Examples include industrial robots, autonomous mobile robots, drones, laboratory systems, vehicles, and emerging humanoid platforms.

The difference is consequence. A poor sentence can be edited. A poor physical action may damage equipment, interrupt production, or threaten safety. Physical AI therefore requires stronger testing, simulation, redundancy, monitoring, and operational boundaries.

Robotics is already infrastructure

The International Federation of Robotics reported 542,000 industrial robot installations in 2024 and an operational global stock of approximately 4.664 million units. Industrial deployment is not a distant scenario; it is a mature base upon which more capable perception, language interfaces, and adaptive control are being added. [7]

Why the edge matters

Edge computing places computation near the machine or sensor. This can reduce latency, preserve operation when connectivity fails, and limit the amount of data sent elsewhere. Cloud systems remain valuable for large-scale training, fleet analytics, and global coordination. Many physical AI systems will combine both.

The architectural question is continuous: what must be decided locally, what can be deferred, and what requires coordination across the fleet?

The intermediate layer: fleet operations

A single robot is a product. A thousand robots create an infrastructure problem. Fleets require identity, task allocation, software updates, telemetry, maintenance, charging, cybersecurity, incident response, and rules for human override.

This is why terms such as robot fleet orchestration, physical AI operating systems, control planes, machine identity networks, and autonomous workforce platforms are likely to become more important. The machine attracts attention; the coordination layer determines whether deployment can scale.

INTERMEDIATE MODEL

A simple model of physical AI

Sense → interpret → decide → act → observe → learn. Safety and control must surround the entire loop, not be added only at the moment of action.

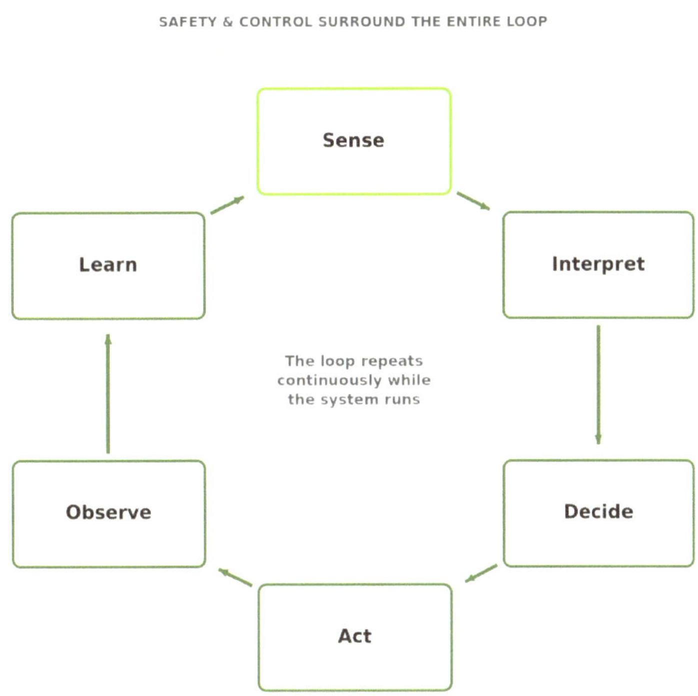

CHAPTER 13

Financial Infrastructure and Machine Commerce

AI does not only analyze markets. It increasingly participates in the systems through which money, obligations, and risk move.

Finance is coordination infrastructure

Payments, exchanges, clearing, custody, settlement, treasury, liquidity, identity, and compliance are systems for coordinating economic activity. Most users see the product interface; the infrastructure determines whether a transaction becomes final, who bears risk, and how capital moves.

The United States moved most securities transactions to a T+1 settlement cycle on May 28, 2024. The change illustrates how apparently technical market plumbing can reduce time and risk across a large system. Shorter cycles require process, technology, and coordination—not merely a faster button. [8]

Everyday AI in finance

Individuals and businesses already use AI to categorize expenses, explain statements, draft forecasts, identify anomalies, prepare scenarios, and translate financial language. These are useful applications when the source data is controlled and the calculations are checked.

The risk is authority drift. A tool that explains a financial concept may be treated as an adviser. A model that finds a pattern may be treated as proof. Maintain a clear boundary between assistance, analysis, and regulated or fiduciary judgment.

The intermediate layer: agents as economic participants

As agents gain permission to purchase services, allocate budgets, negotiate, or trigger payments, financial infrastructure must identify the agent, verify authority, enforce limits, and settle the transaction. The machine may act quickly, but accountability must remain legible.

This creates new categories around agent payments, machine identity, programmable treasury, autonomous procurement, settlement layers, and capital routing. The future of agentic commerce depends less on whether a model can click 'buy' than on whether institutions can trust, constrain, and reconcile the action.

MARKET PRINCIPLE

Machine-speed finance needs human-readable control

Faster execution increases the value of limits, logs, approvals, exception handling, and settlement certainty. Automation does not remove governance; it compresses the time available to exercise it.

CHAPTER 14

Sovereign AI and Strategic Compute

Nations increasingly view compute, energy, data, talent, and infrastructure as components of economic and security capacity.

From commercial service to strategic resource

For much of the cloud era, compute was treated primarily as a commercial service. AI is changing the political meaning of capacity. Advanced systems influence research, defense, public services, industrial productivity, communications, and economic competitiveness. Dependence on external providers therefore becomes a strategic question.

Sovereign AI does not necessarily mean that every country builds every component. It means that governments and institutions evaluate where critical dependencies exist and what level of control, resilience, legal jurisdiction, or domestic capacity is required.

The sovereign stack

- Compute: accelerators, supercomputers, data centers, cloud capacity, and scheduling.
- Energy: reliable generation, transmission, cooling, and local grid capability.
- Data: lawful access, quality, language coverage, security, and governance.
- Talent: researchers, engineers, operators, domain experts, and public-sector capability.
- Institutions: procurement, standards, safety, competition policy, and public accountability.

A nation can possess strong models and weak infrastructure, or abundant infrastructure and weak adoption. Strategic capacity depends on the system.

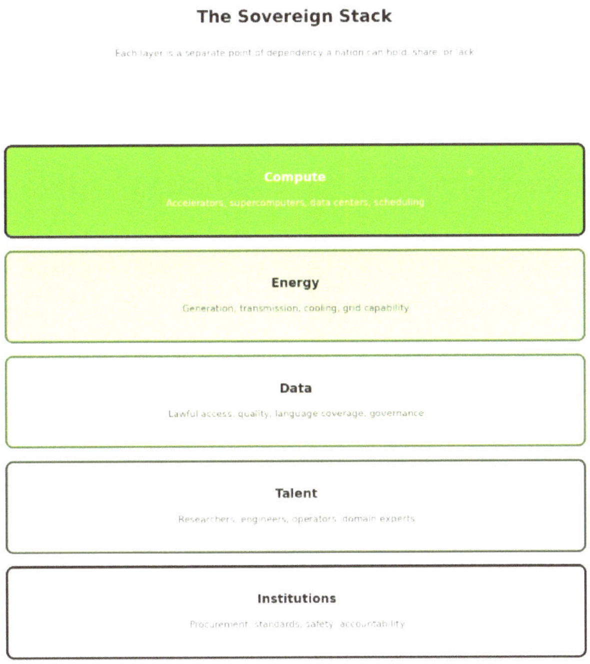

Figure: the sovereign stack — compute, energy, data, talent, and institutions, each a separate point of national dependency.

Present-day signals

The European Union's AI Factories program uses EuroHPC supercomputing capacity to support advanced AI development, while its broader AI Continent strategy emphasizes computing infrastructure, data access, adoption, and skills. In the United States, the Department of Energy has identified federal sites and energy infrastructure opportunities for AI data-center development. These initiatives differ in structure, but both treat compute and power as policy concerns rather than ordinary software procurement. [9] [10]

The intermediate layer: compute markets and public capacity

Once compute is treated as strategic, allocation becomes political and economic. Who receives scarce capacity? At what price? Should public research institutions receive reserved access? How should cross-border workloads be governed? What happens during emergencies or supply disruptions?

These questions point toward national compute exchanges, capacity reservation systems, sovereign cloud frameworks, shared research infrastructure, and new forms of public-private partnership. The eventual institutions remain uncertain, but the category is forming.

STRATEGIC LENS

Sovereignty is a dependency question

The practical question is not whether a system is completely domestic. It is whether critical dependencies are understood, diversified, governed, and resilient enough for the intended use.

CHAPTER 15

Orbital, Edge, and Networked Intelligence

AI increasingly depends on systems that observe, communicate, navigate, and coordinate beyond centralized data centers.

Infrastructure extends beyond the cloud

Satellites support communications, navigation, timing, Earth observation, weather analysis, defense, logistics, agriculture, and disaster response. Edge devices bring intelligence into vehicles, factories, phones, sensors, and remote environments. Together they broaden the geography of computation.

The cloud remains central, but many decisions cannot wait for a distant data center. Latency, connectivity, bandwidth, privacy, and resilience push parts of the AI stack outward.

Orbital systems as operational infrastructure

The important shift is from isolated spacecraft to networks. Constellations, ground stations, data pipelines, traffic coordination, cybersecurity, and service integration create an operational layer. The satellite itself is only one component.

As orbital activity grows, the coordination problems become familiar: identity, scheduling, spectrum, collision avoidance, data rights, maintenance, interoperability, and governance. Space becomes an infrastructure environment rather than only a destination.

The intermediate layer: distributed intelligence

Distributed systems divide work across locations. A sensor may detect an event locally, a satellite may relay data, an edge device may classify it, and a cloud platform may coordinate the wider response. The design must account for delays, failures, conflicting observations, and security across the network.

This connects orbital infrastructure to the same core concepts explored throughout the book: context, permissions, trust, allocation, observability, and control planes. Different environments create different constraints, but the coordination problem persists.

BYEGIG LENS

The system-of-systems view

AI, communications, positioning, sensing, energy, and physical operations increasingly reinforce one another. The future infrastructure map is not a collection of separate sectors; it is a network of interdependent systems.

CHAPTER 16

Categories, Language, and the Markets AI Creates

New markets become easier to build when participants develop shared language for the problem, the system, and the value being created.

Language is a coordination tool

Cloud computing, cybersecurity, software as a service, digital identity, and data centers now feel like established categories. Each began as a collection of technical capabilities, competing terms, and uncertain business models. Language stabilized as participants learned what the category contained and why it mattered.

A category lets founders explain what they build, customers understand what they buy, investors compare opportunities, analysts measure a market, and policymakers determine what may require oversight. Naming does not create the underlying need, but it helps organize activity around it.

Why infrastructure language is different

Consumer brands often favor brevity and emotion. Infrastructure terminology frequently favors clarity, function, legitimacy, and scale. Control plane, settlement network, identity infrastructure, capacity market, operating system, and orchestration layer are descriptive because the buyer needs to understand the role the system plays.

This is the strategic logic behind ByeGig's focus on category-defining digital assets. The objective is not to collect fashionable phrases. It is to identify recurring infrastructure problems before the market converges on the language used to describe them.

The intermediate layer: category architecture

A mature category contains more than a name. It develops subcategories, standards, buyer roles, metrics, institutions, and adjacent markets. Compute markets may create exchanges, brokers, capacity products, settlement systems, risk tools, and policy frameworks. Physical AI may create fleet systems, machine identity, charging networks, safety standards, and labor platforms.

Thinking in category architecture helps readers understand why emerging technologies create second-order opportunities. The visible product is often the beginning. Infrastructure, coordination, trust, finance, and governance expand around it.

COPY-AND-USE PROMPT

Map an emerging category

Analyze the emerging field of [field]. Separate current products from enabling infrastructure. Identify: core resource, participants, coordination problem, trust problem, likely market mechanisms, regulatory questions, and terms that appear to be stabilizing. Mark which conclusions are supported by present evidence and which are speculative.

BYEGIG PRINCIPLE

Before markets exist, someone must name them

Technology creates possibility, infrastructure creates scale, and language creates coordination. Durable categories emerge once all three begin to reinforce one another.

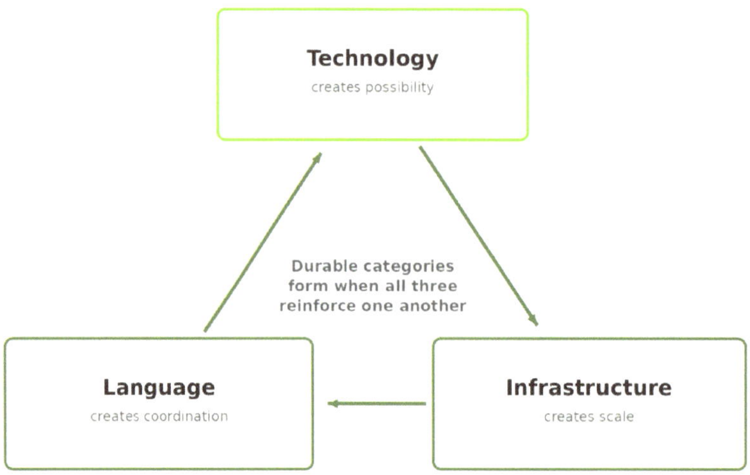

How These Systems Connect

Part II moved through seven infrastructure layers — compute and energy, trust and verification, physical AI, financial infrastructure, sovereign AI, orbital and edge networks, and the categories and language that organize adoption. Each is a distinct system, but none stands alone. The table below is a quick reference for what each layer provides, what it depends on, and where in this book to return to it.

System	What It Provides	Key Dependency	Chapter
Compute & Energy	Raw processing power and the electricity to run it	Power capacity, chip supply	Ch. 10
Trust & Verification	Confidence that content and identity are genuine	Provenance and identity standards	Ch. 11
Physical AI & Robotics	AI acting and sensing in the physical world	Sensors, actuators, edge compute	Ch. 12
Financial Infrastructure	Machines transacting and settling value directly	Machine-readable payment rails	Ch. 13
Sovereign AI	National control over AI capacity	Compute, energy, data, talent, institutions	Ch. 14
Orbital & Edge Networks	Intelligence distributed beyond data centers	Satellite links, local compute	Ch. 15
Categories & Language	Markets and vocabulary that organize adoption	Naming, standards, shared terms	Ch. 16

This chapter treated each layer as a short introduction. For a longer, category-by-category look at the same ground — plus two categories this book only mentions in passing, the drone economy and the premium domain name market — see ByeGig's *Field Guide to Emerging Infrastructure*.

Part III turns from these systems to practice — how a person or an organization builds the habits, permissions, and adoption plan that put all of it to use.

Part III

Building an AI Practice

The objective is not to use AI everywhere. It is to build a system that produces measurable value with controlled risk.

CHAPTER 17

A Personal and Organizational Adoption System

Move from experimentation to a portfolio of verified workflows, then scale only what deserves to scale.

Start with work, not tools

The AI market changes too quickly for a durable strategy to begin with a list of products. Begin with work: the recurring tasks, bottlenecks, decisions, information flows, and quality problems that define your day or organization.

OpenAI's enterprise guidance recommends identifying concrete opportunities, teaching foundational use cases, and prioritizing high-impact, lower-effort applications before attempting complex transformation. The general principle is vendor-neutral: adoption succeeds when capability is attached to an observable problem. [11]

Create an AI workflow portfolio

Maintain a simple register with the workflow name, owner, purpose, inputs, outputs, time saved, quality effect, risk level, review method, and current status. This prevents promising experiments from disappearing and weak experiments from quietly becoming permanent.

Classify workflows into four groups: personal assistance, team production, decision support, and autonomous execution. The governance burden rises across the groups.

Use risk tiers

- Low risk: brainstorming, formatting, learning, drafting from approved material, and reversible personal organization.
- Moderate risk: customer-facing content, operational analysis, internal recommendations, or work involving confidential information.

- High risk: legal, medical, financial, safety, employment, security, regulatory, or autonomous actions with meaningful consequences.

Risk is determined by both task and deployment. A summary may be low risk when used privately and high risk when it determines whether a customer, employee, or patient receives a service.

Govern, map, measure, manage

NIST's AI Risk Management Framework organizes responsible practice around four functions: Govern, Map, Measure, and Manage. In plain language: establish ownership and policy; understand the use case and affected people; test performance and risk; then operate, monitor, and improve the system. [12]

This framework is useful even for a small team: governance can be a one-page rule set, mapping a workflow diagram, measurement a test set and error log, and management nothing more than a named owner, a review cadence, and a shutdown procedure. [12]

Measure what matters

Track time saved, but do not stop there. Measure error rate, revision burden, turnaround time, user satisfaction, consistency, and whether the workflow expands the range of work a person can perform. Productivity may appear as speed, quality, capacity, or reduced cognitive friction.

OpenAI's 2025 enterprise report found that surveyed workers in several functions reported meaningful time savings and operational improvements, while heavier users reported larger gains. Vendor research should be interpreted with appropriate caution, but the broader lesson is practical: value depends on depth of integration and repeated use, not one-off demonstrations. [13]

The intermediate layer: change management

AI adoption changes roles and expectations. Employees need permission to experiment, boundaries for safe use, examples relevant to their work, and a way to share effective practices. Leaders need to distinguish genuine productivity from hidden rework and quality risk.

The strongest adoption systems combine top-down clarity with bottom-up discovery. Leadership defines goals and controls. Users identify the real workflows. Technical and domain experts turn the best experiments into dependable systems.

IMPLEMENTATION MODEL

The adoption equation

Useful capability × relevant context × repeated practice × trustworthy controls = durable value. A weakness in any factor limits the whole system.

APPENDIX A

A 30-Day AI Adoption Plan

A practical sequence for building confidence without automating too much too soon.

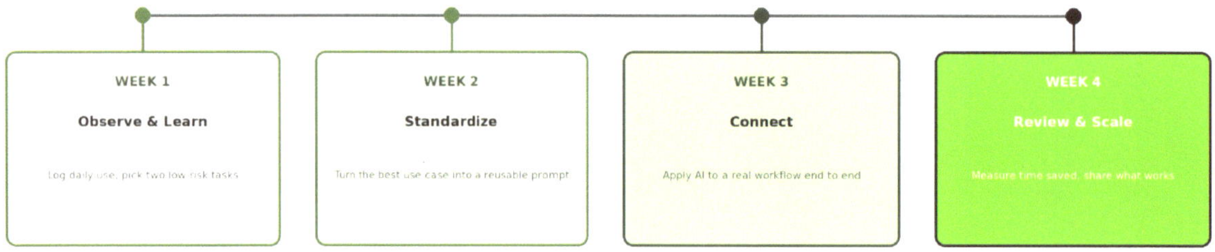

Figure: the 30-day adoption plan at a glance — observe, standardize, connect, then review and scale.

Week 1 — Observe and learn

- List twenty recurring tasks and mark those involving drafting, summarizing, comparison, explanation, classification, or planning.
- Choose two low-risk tasks and use AI for them at least three times.
- Create a private note titled “AI Workflow Log” and record prompts, results, failures, and time saved.
- Practice separating facts, assumptions, and recommendations.

Week 2 — Standardize

- Turn the strongest use case into a reusable prompt with objective, context, inputs, constraints, output, and evaluation.
- Create one project brief or context document for a recurring area of work.
- Add a verification checklist for any task involving external facts.
- Test the workflow on an ordinary case and an unusual case.

Week 3 — Connect

- Use AI with a real document, spreadsheet, or dataset.
- Break one multi-step process into a workflow map.

- Identify which step could be automated safely and which requires human approval.
- Create a small evaluation set containing five examples of acceptable and unacceptable output.

Week 4 — Review and scale

- Calculate approximate time saved and revision burden.
- Delete or redesign workflows that create hidden rework.
- Share one proven workflow with a colleague or document it for future reuse.
- Choose the next subject to deepen: research, data, agents, compute, trust, robotics, finance, or governance.

REFLECTION

End-of-month question

Which AI workflow would you continue using even if the novelty disappeared? That is the workflow most likely to create durable value.

APPENDIX B

Reusable Prompt Patterns

Templates are starting points. Adapt the context, risk controls, and evaluation criteria to the task.

COPY-AND-USE PROMPT

Explain at three levels

Explain [topic] at three levels: (1) a concise everyday explanation, (2) a professional explanation for a technology-literate reader, and (3) an intermediate explanation covering architecture, trade-offs, and common failure modes. End with five questions that test whether I understood it.

COPY-AND-USE PROMPT

Turn notes into a decision brief

Using only the notes below, prepare a decision brief with: objective, known facts, assumptions, options, trade-offs, recommendation, what would change the recommendation, and next actions. Mark missing information explicitly. Notes: [paste]

COPY-AND-USE PROMPT

Compare sources

Compare the attached sources. For each major claim, show which sources support it, which contradict it, and whether the disagreement is factual, methodological, or interpretive. Do not average incompatible claims. Identify the strongest primary source.

COPY-AND-USE PROMPT

Prepare for a meeting

Prepare me for a meeting with [person/organization] about [topic]. Using the materials provided, summarize their likely priorities, unresolved issues, questions I should ask, claims I should verify, and the outcome I should seek. Separate evidence from inference.

COPY-AND-USE PROMPT

Improve a workflow

Analyze the workflow below for delay, duplication, manual transfer, unclear ownership, and error risk. Recommend improvements in three levels: no automation, AI assistance, and bounded automation. State the controls required at each level.

COPY-AND-USE PROMPT

Challenge a plan

Assume the plan fails twelve months from now. Produce a pre-mortem identifying the most plausible causes, early warning indicators, and mitigations. Prioritize causes that are both consequential and controllable.

COPY-AND-USE PROMPT

Create an evaluation rubric

Create a scoring rubric for [task]. Include criteria, definitions, examples of strong and weak performance, disqualifying errors, and a minimum acceptable threshold. The rubric must be usable by two independent reviewers.

COPY-AND-USE PROMPT

Build a learning path

Design a four-week learning path for [topic] for a technology-literate beginner. Each week should include concepts, one practical exercise, one intermediate concept, a verification question, and a small deliverable.

APPENDIX C

Glossary

Plain-language definitions for the concepts used throughout the book.

Agent: An AI-enabled system that can pursue an objective across multiple steps, use tools, observe results, and decide what to do next within defined permissions.

Artificial intelligence: A broad field concerned with systems that perform tasks associated with perception, language, prediction, reasoning, learning, or action.

Context: The information available to a model during a particular interaction.

Context window: The amount of information a model can consider at one time.

Control plane: A coordinating layer that manages policy, routing, allocation, permissions, and system behavior.

Embedding: A numerical representation that helps systems compare meaning or similarity.

Evaluation: A test or measurement process used to determine whether an AI system performs acceptably for a defined purpose.

Fine-tuning: Further training used to adapt a model's behavior to particular examples or tasks.

Grounding: Constraining an answer to supplied sources, tools, data, or rules.

Hallucination: A fluent but unsupported or incorrect output presented as though it were reliable.

Inference: The process of using a trained model to produce an output.

Large language model: A model trained on large amounts of language data to predict and generate text and related outputs.

Machine identity: The credentials, ownership, permissions, and attributes used to identify a software agent, device, service, or robot.

Model: A trained computational system that transforms inputs into predictions or generated outputs.

Multimodal: Able to work across more than one form of information, such as text, images, audio, or video.

Orchestration: The coordination of multiple models, agents, tools, resources, or workflow steps.

Physical AI: AI connected to systems that sense and act in the physical world.

Prompt: The instruction and context provided to an AI model.

Provenance: Information about the origin and modification history of content or data.

Retrieval-augmented generation: A method in which a system retrieves relevant information and supplies it to a model while generating an answer.

Sovereign AI: The capacity of a nation or institution to govern critical AI dependencies such as compute, data, energy, talent, and legal jurisdiction.

Tool use: The ability of an AI system to call an external capability such as search, code execution, a database, or another application.

Workflow: A defined sequence of tasks, decisions, inputs, and outputs used to accomplish a recurring objective.

BACK MATTER

Notes and Sources

Selected primary and institutional references used in this edition. Accessed July 2026.

1. Stanford Institute for Human-Centered Artificial Intelligence, The 2026 AI Index Report. https://hai.stanford.edu/ai-index/2026-ai-index-report
2. OECD, "AI use by individuals surges across the OECD as adoption by firms continues to expand," January 28, 2026. https://www.oecd.org/en/about/news/announcements/2026/01/ai-use-by-individuals-surges-across-the-oecd-as-adoption-by-firms-continues-to-expand.html
3. OpenAI, "ChatGPT usage and adoption patterns at work," January 22, 2026. https://openai.com/business/guides-and-resources/chatgpt-usage-and-adoption-patterns-at-work/
4. OpenAI Help Center, "Prompt engineering best practices for ChatGPT," updated 2026. https://help.openai.com/en/articles/10032626-prompt-engineering-best-practices-for-chatgpt
5. International Energy Agency, Energy and AI: Energy demand from AI. https://www.iea.org/reports/energy-and-ai/energy-demand-from-ai
6. Coalition for Content Provenance and Authenticity, C2PA Technical Specification 2.4 and explainer, April 2026. https://spec.c2pa.org/specifications/specifications/2.4/index.html
7. International Federation of Robotics, World Robotics 2025 industrial robot statistics. https://ifr.org/ifr-press-releases/news/global-robot-demand-in-factories-doubles-over-10-years
8. U.S. Securities and Exchange Commission, statement on implementation of the T+1 settlement cycle, May 21, 2024. https://www.sec.gov/newsroom/press-releases/2024-62
9. European Commission, "AI Factories," updated April 23, 2026. https://digital-strategy.ec.europa.eu/en/policies/ai-factories
10. U.S. Department of Energy, "DOE Identifies 16 Federal Sites Across the Country for Data Center and AI Infrastructure," April 3, 2025. https://www.energy.gov/articles/doe-identifies-16-federal-sites-across-country-data-center-and-ai-infrastructure
11. OpenAI, "Identifying and scaling AI use cases," 2026. https://openai.com/business/guides-and-resources/identifying-and-scaling-ai-use-cases/

12. National Institute of Standards and Technology, AI Risk Management Framework and Generative AI Profile. https://www.nist.gov/itl/ai-risk-management-framework

13. OpenAI, The State of Enterprise AI 2025 Report. https://openai.com/business/guides-and-resources/the-state-of-enterprise-ai-2025-report/

What's next for this book

This edition establishes the book's practical and conceptual foundation. Future editions may add expanded case studies, diagrams, sector-specific chapters, exercises, interviews, original charts, and a deeper treatment of compute markets, defense infrastructure, world models, energy systems, and category ownership.

BACK MATTER

About ByeGig

ByeGig acquires and develops premium domain names in emerging infrastructure categories—artificial intelligence, compute, robotics and physical AI, defense and counter-drone systems, energy, orbital and space systems, financial infrastructure, and trust and identity verification. This book grew out of the same research practice that drives ByeGig's portfolio decisions: tracking how technologies mature into infrastructure, and how infrastructure eventually needs a name.

Readers who want the deeper argument behind this book's infrastructure chapters can find it in ByeGig's companion library:

- **The Emerging Infrastructure Era —** the full essay collection behind the infrastructure chapters in this book, covering AI, compute, sovereign technology, robotics, defense, finance, trust, energy, and orbital systems in depth.
- **The Category Ownership Playbook —** a practical guide to recognizing, naming, and claiming an emerging technology category, expanding on Chapter 16's discussion of language and markets.
- **The Field Guide to Emerging Infrastructure —** a sourced, category-by-category reference covering the same ten infrastructure categories introduced across Part II of this book.
- **Own the Layer —** a short, primer-length version of the same argument, for readers who want the thesis without the workbook.
- **Working Through the AI Shift —** a companion volume on jobs, skills, and career resilience, for readers who want to understand what this technology means for their own work, not just how to use the tools.

Readers interested in the underlying domain portfolio or further writing on the infrastructure era can follow ByeGig's continuing coverage through its Substack and LinkedIn channels.

www.ingramcontent.com/pod-product-compliance
Lightning Source LLC
LaVergne TN
LVHW072011300926
849160LV00003B/17